S0-EJW-622

At the Frog Pond

At the Frog Pond

by Tilde Michels
with pictures by Reinhard Michl

Translated from the German
by Nina Ignatowicz

J. B. LIPPINCOTT NEW YORK

Did you ever wonder
how a tadpole turns into a frog?

Did you ever stumble onto a secluded spot
where you could hear and see
the wondrous ways of nature—
a clearing,
a marsh—
or a small frog pond?

There the water shimmers blue,
or gray, or brown, or green—
just like the trees and sky and clouds
that are mirrored in the pond's surface.

In the first warm days of spring
signs of life are everywhere.
Creatures wake from their winter sleep.
New nests are begun.
Muskrats come out of their burrows.

Frogs swim to the surface and mate.
Wobbly clusters of frog eggs
drift to the pond's banks,
where they hatch into thousands of tadpoles.

But not all the tadpoles survive.
Many are eaten by other creatures,
who need them for food.

Toads too come to the pond to spawn.
They migrate great distances
to lay their eggs in the same pond
where they were born.
There are many different kinds
of frogs and toads,
but they all go through the same changes—
from egg to tadpole,
from tadpole to full-grown frog or toad.

On clear nights the moon glimmers on the pond.

Sounds of life are all around.

There are squeals and gurgles,

chirps and burbles,

and rustlings in the reeds.
A shadow glides over the meadows—
an owl searching for prey.

At dawn a heron comes to the edge of the pond
to look for fish in the shallow waters.
He wades into the water and waits there,
motionless.
His eyes are alert to the slightest movement.
Suddenly he darts out his head
and plunges it into the water.
A moment later he comes up with a fish in his beak.

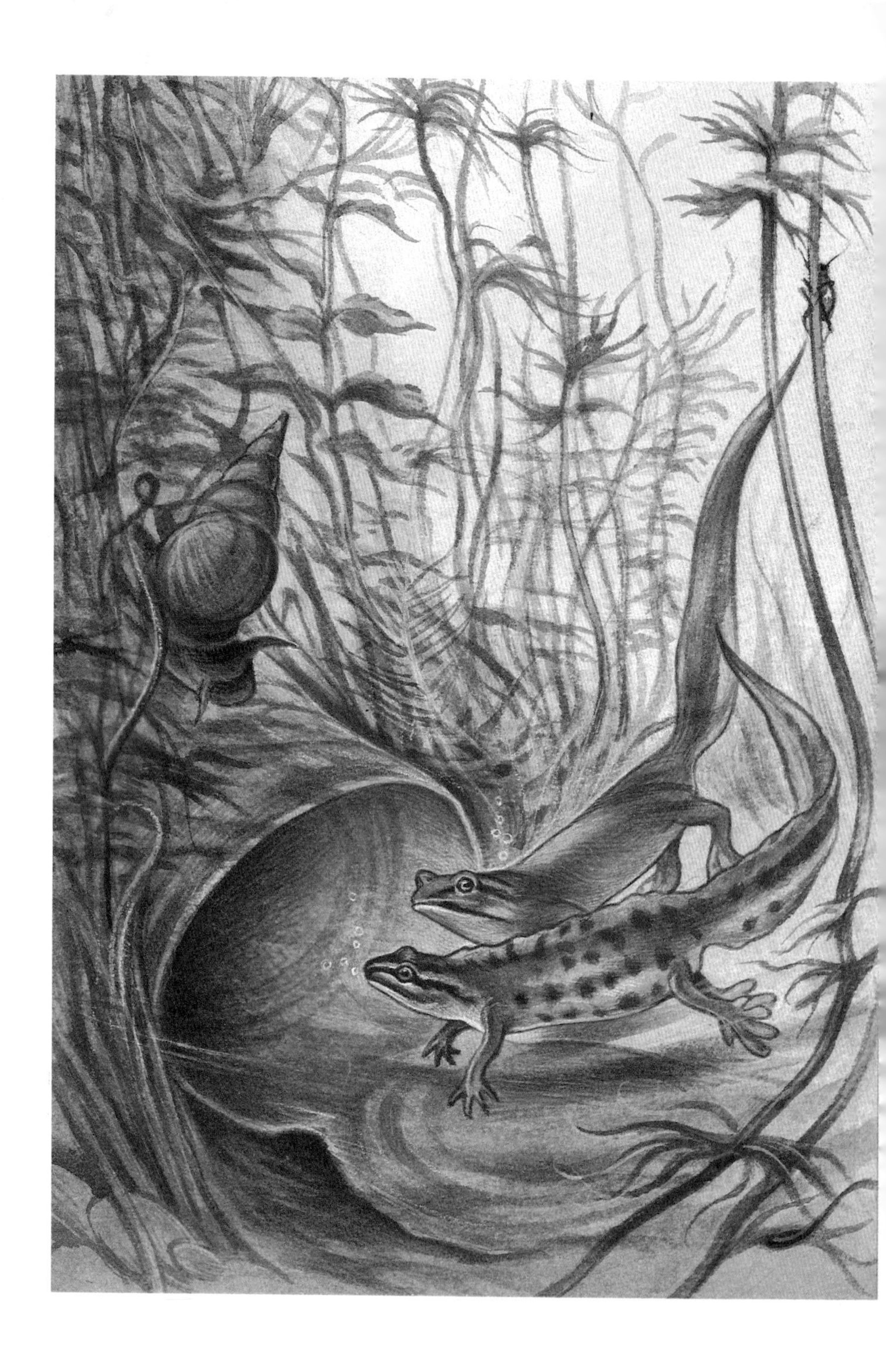

The pike too is always hunting for food.
He is a thief and a glutton
who snaps at anything
that passes by his mouth.
Small fish scoot away from him.
Pond salamanders scuttle into the mud
as soon as they see him.

When water lilies are in bloom,
it is summer.
At noon a calm settles over the little pond.
All sounds cease.
The sun hangs high up in the sky,
its light glittering over the water.
The heat vibrates.
Not a blade stirs.
The reeds stand motionless.

All at once the air becomes heavy and humid.
Steel-gray clouds start piling up.
Thunder rumbles in the distance.
A storm is brewing.
Soon the first fat drops of rain strike the pond,
burst, and make bubbles and circles
that dance over the water.
Then the storm unleashes its fury.
Lightning forks down from the clouds.
The rain pelts the pond,
and the wind lashes the grass on the banks.

All the animals have hidden from the storm.
The birds have fled to the trees,
the insects have crawled into the moss or under leaves.
Frogs have burrowed into the muddy ground.

As suddenly as it appeared, the storm disappears.
Mist rises from the marsh,

the pond shimmers in the twilight,

and all the frogs begin a grand frog concert.

Sometimes a pair of ducks will visit,
swim around the duckweeds for a while,
then leave again.

A solitary hawk circles the pond.
Even from way up high, he can see
anything scurrying around on the ground.

He has hungry fledglings in his nest.
He is looking for food for them.

If you are one of the lucky ones
to find such a spot,
walk over quietly,
sit down on a sun-warmed log,
and look and listen.
And for a little while
you will be a part of things
that live and grow
around this small frog pond.

At the Frog Pond

Originally published in Germany under the title *Am Froschweiher* by
Deutscher Taschenbuch Verlag, Munich

Typography by Carol Barr
1 2 3 4 5 6 7 8 9 10
First American Edition

Library of Congress Cataloging-in-Publication Data
Michels, Tilde.
At the frog pond.

Translation of: Am Froschweiher.
Summary: Describes the animal life at a secluded pond during spring and summer days.
1. Pond fauna—Juvenile literature. 2. Pond ecology—Juvenile literature. 3. Spring—Juvenile literature. 4. Summer—Juvenile literature. [1. Pond animals] I. Michl, Reinhard, ill. II. Title.
QL146.3.M5313 1989 591.52′6322 87-37835
ISBN 0-397-32314-X
ISBN 0-397-32315-8 (lib. bdg.)